Summary and Analysis of

THE GENE

An Intimate History

Based on the Book
by Siddhartha Mukherjee

WORTH BOOKS
SMART SUMMARIES

All rights reserved, including without limitation the right to reproduce this book or any portion thereof in any form or by any means, whether electronic or mechanical, now known or hereinafter invented, without the express written permission of the publisher.

This Worth Books book is based on the 2016 ebook edition of *The Gene* by Siddhartha Mukherjee, published by Scribner.

Summary and analysis copyright © 2017 by Open Road Integrated Media, Inc.

ISBN: 978-1-5040-4669-5

Worth Books
180 Maiden Lane
Suite 8A
New York, NY 10038
www.worthbooks.com

Worth Books is a division of Open Road Integrated Media, Inc.

The summary and analysis in this book are meant to complement your reading experience and bring you closer to a great work of nonfiction. This book is not intended as a substitute for the work that it summarizes and analyzes, and it is not authorized, approved, licensed, or endorsed by the work's author or publisher. Worth Books makes no representations or warranties with respect to the accuracy or completeness of the contents of this book.

Contents

Context	1
Overview	3
Summary	7
Timeline	31
Cast of Characters	35
Direct Quotes and Analysis	39
Trivia	43
What's That Word?	47
Critical Response	51
About Siddhartha Mukherjee	53
For Your Information	55
Bibliography	57

Context

The Gene was published in May 2016, thirteen years after the mapping of the human genome was completed and only a year after the announcement that an efficient method for modifying human genes had been discovered. Siddhartha Mukherjee, the Pulitzer Prize-winning author of *The Emperor of All Maladies,* describes the evolution of genetic science from its infancy in the late 19th century up to the present day, when the idea that many human traits or characteristics—including sexual preference and intelligence—are largely genetic in nature has taken hold in public discourse.

The larger questions center on the very real possibility of modifying the human genome. Now that

SUMMARY AND ANALYSIS

the science has advanced to the point where this is possible, who should make the choice as to whether it should be done, if at all? Will gene modification be used to eliminate diseases, or for some less noble purpose? The 20th century has already provided lessons on how genetic science can be used to justify atrocities. Today the gene is, more than ever, a "dangerous idea," and how it is put to use will have an enormous influence on humanity's future.

Overview

The Gene: An Intimate History is a rigorously scientific, broadly historical, and candidly personal account of the development of the science of genetics, of the dramatic ways genes can affect us, and of the enormous moral questions posed by our ability to manipulate them.

Siddhartha Mukherjee begins by explaining how three members of his family were unraveled by genetic mental illness. He uses this story to introduce the gene itself and the history of genetics. Like the byte in digital data or the atom in matter, the gene is the most basic unit of information in biology, the lowest common denominator of living organisms. Understanding it is essential to understanding human biology, illness, wellness, temperament, and personality. And, as the author states, the centuries-old quest to control the gene has caused it to become "one of the most powerful and dangerous ideas in the history of science."

The book's structure is historical and chronological. It begins in 350 BC with Aristotle's surprisingly accurate theories of heredity, and ends in 2015, when the technology to modify the genome of a human embryo may be only a few experiments away. Across this vast stretch of time, *The Gene* recounts humanity's fascination with how human traits are inherited, through the stories of the scientists who deciphered the mystery: Gregor Mendel, the Augustinian monk whose research on pea plants in his abbey garden provided the first glimpse of heredity; Charles Darwin, the onetime theology student who advanced the theory of evolution by natural selection; and, in our time, the men and women who took on the immense, daunting task of mapping the entire human genome. The story progresses chapter by chapter like a scientific whodunit, in which each new discovery answers one question but raises another.

Apart from explaining the science of the gene, Mukherjee chronicles its effects on society—not as a biological entity, but as an idea. The emergence of genetics in the early 20th century spawned the eugenics movement, which sought to improve the human race through selective breeding. This notion was later put to horrific use in Nazi Germany, where genetic experiments were performed on concentration camp prisoners, and in the Soviet Union, where brainwashing and "re-education" programs were carried out on dissidents.

More sobering, if less well known, is the story of the eugenics movement in the United States, where thousands of forced sterilizations were performed in the 1920s and '30s on women diagnosed as "feebleminded."

Although much of the content of *The Gene* is highly technical, the explanations are informed by analogies, some of them folksy, others contemporary, all of them engaging. To describe how, even today, medicines are able to affect only a tiny fraction of the human body's molecules, Mukherjee tells us: "If human physiology is visualized as a vast global telephone network . . . medicinal chemistry is a pole operator in Wichita tinkering with a few lines in the network's corner."

As *The Gene* traces its chronological arc, Mukherjee returns time and again to the story of his family. He speaks frankly, almost clinically, about the mental illnesses affecting his uncles and cousin and the genetic ailment suffered by his father (normal pressure hydrocephalus, or NPH). Nor does he spare himself from examination; the realization that these traits could be passed on leads him to speculate upon—and

SUMMARY AND ANALYSIS

even calculate, mathematically—the chances that he or his daughters might be likewise affected.

In another context, this might seem self-indulgent. But *The Gene* is not the "intimate history" of the author alone. Anyone who reads it will inevitably wonder about their own inherited traits of biology and personality, the illnesses in their family, and the possibilities offered by gene therapy. Perhaps more importantly, they will also reflect on the huge ethical questions that arise from our knowledge of the tiniest unit of information in the human body.

Summary

Prologue: Families

Mukherjee's uncles Rajesh and Jagu and his cousin Moni have all suffered from mental illnesses with genetic origins. Their stories serve to introduce the gene, the "fundamental unit of heredity" that holds the key to the physical and emotional characteristics of each human life.

Need to Know: No understanding of human biology—physiology or behavior—can be complete without first understanding the nature of the gene.

SUMMARY AND ANALYSIS

Part One: The "Missing Science of Heredity"

The Discovery and Rediscovery of Genes (1865-1935)

The Walled Garden

The concept of the gene was first formed by Gregor Johan Mendel in 1865. However, theories of how human traits are transmitted date back to Aristotle, who theorized that they were contained in "instructions" within the body that were passed on from generation to generation.

Aristotle's theories of heredity were surprisingly accurate; but after him, little progress was made in genetics until the 19th century.

"The Mystery of Mysteries"

Charles Darwin's 1831–1836 voyage to South America and the Pacific to collect animal specimens and fossils of their predecessors led him to believe that new species were created in a struggle for survival. The ones best equipped to adapt to shifting conditions would endure, in a process of "natural selection."

Need to Know: Darwin's *On the Origin of Species*

proposed the theory of evolution but did not provide an explanation for how heredity works.

The "Very Wide Blank"

Darwin later sought to explain heredity with the theory of "pangenesis." He believed that tiny particles called *gemmules*, contained within the cells of living organisms, passed on their essential traits to their offspring, blending the characteristics of both mother and father. His colleagues dismissed the theory because it did not explain why some traits disappeared and then reappeared in later generations.

Need to Know: Darwin's failure to explain heredity came about in part because he was a brilliant naturalist, but not adept at carrying out experiments.

"Flowers He Loved"

Gregor Mendel's research breeding hybrid pea plants led him to conclude that the traits of organisms must be contained in single, indivisible units. Although he did not invent the term, Mendel had identified the fundamental qualities of what we call a gene today.

Need to Know: Mendel's seminal work on heredity

was received with silence; his contribution to science was not recognized until long after his death in 1884.

"A Certain Mendel"

The Dutch botanist and geneticist Hugo de Vries, aware of Mendel's work, observed that plant species spontaneously generated new varieties—ones with longer stems or flowers of different colors. He termed these variants *mutants* and the process, *mutation*. Nature, he posited, spontaneously created mutants with specific qualities and the ones best adapted to changing circumstances were able to survive, creating new species.

Need to Know: Hugo de Vries's discovery of spontaneous mutation extended Darwin's theory of natural selection.

Eugenics

Darwin's cousin Francis Galton was an early proponent of eugenics, which advocated the selective breeding of men and women in order to create a superior society. The notion appealed to a British ruling class appalled over the rising power and numbers of the working class, and to sectors in the United States concerned about the effects of massive immigration.

Need to Know: The eugenics movement in the United States and Britain used theories of genetics to justify social experiments that included forced sterilization.

"Three Generations of Imbeciles Is Enough"

During the 1920s, individual states began to pass laws authorizing imprisonment or sterilization of men and women deemed to have genetic criminal tendencies or mental deficiencies. The first legally forced sterilization was performed on Carrie Buck, an impoverished single mother in rural Virginia who, like her own mother, was confined to a penal colony after being diagnosed as a "moron."

Buck, who did not express any objection to the procedure, was sterilized under a 1924 Virginia law that was later upheld by the US Supreme Court, opening the way for thousands of such forced sterilizations.

Need to Know: In the United States, eugenics became, if not the law of the land, a powerful tool for people who sought to protect society from what they regarded as moral and social decay.

SUMMARY AND ANALYSIS

Part Two: "In the Sum of the Parts, There Are Only the Parts"

Deciphering the Mechanism of Inheritance (1930-1970)

"Abhed"

Thomas Morgan, a researcher at Columbia University in New York, conducted extensive experiments on fruit flies and found that genes were linked to one another and to chromosomes. Based upon this notion of "linkage," a student of Morgan's, Alfred Sturtevant, constructed the first genetic map.

Need to Know: Thomas Morgan's graphic description of genes' location on chromosomes as "beads on a string" illustrated their linkage to one another, and to chromosomes. This was the first genetic map.

Truths and Reconciliations

Theodosius Dobzhansky, a Ukrainian biologist who immigrated to the United States, discovered that while genes (or genotype) determine traits (phenotype), environment also has an effect, as does chance. He exposed groups of fruit flies to different temperature levels over several generations; the resulting changes in their physical traits showed that environment also influences an organism's genetic makeup.

In this way, Dobzhansky also proved that gene selection alone could not be the basis for creating superior beings, as the eugenicists believed.

Need to Know: Dobzhansky's experiments led to the equation "genotype + environment + triggers + chance" as the explanation for physical attributes (phenotype) of living creatures.

Transformation

Frederick Griffith, a British medical officer seeking a vaccination for the Spanish flu, found that two different varieties of bacteria could exchange genes, with one acquiring the attributes of the other. This meant attributes could be transmitted without reproduction, in a process he called "transformation." Hermann Muller proved that genes could be altered by radiation to produce mutant varieties. This meant that genes were actual, physical matter that could be manipulated by man.

Need to Know: The research by Griffith and Muller marked the first step toward manipulation of human genes.

SUMMARY AND ANALYSIS

Lebensunwertes Leben (Lives Unworthy of Living)

The Nazis' mass experiment in "racial hygiene" began with sterilization and quickly moved to the imprisonment and execution of mental defectives, criminals, dissidents, Jews, and Gypsies. The Russian Communists, meanwhile, used theories of gene manipulation to justify a massive program to erase all differences in thought through brainwashing or "re-education."

Need to Know: Both the Nazis in Germany and the Communists in the Soviet Union used the language of genetics, although not exactly the science itself, to justify their atrocities.

"That Stupid Molecule"

In 1944, Oswald Avery, a professor at Rockefeller University in New York, experimenting with virulent and nonvirulent bacteria, proved that the nucleic acid DNA (deoxyribonucleic acid) is the molecule that carries an organism's genetic information. Avery based his discovery on the work done previously by Frederick Griffith.

Need to Know: The identification of DNA (scorned as "that stupid molecule" by one scientist) as the carrier of the genetic code served as a landmark to direct future research in genetics.

"Important Biological Objects Come in Pairs"

Researcher Rosalind Franklin, using an X-ray machine, took the first useful photos of DNA molecules, which permitted her colleagues James Watson and Francis Crick to determine the molecule's structure. After many attempts, they created the iconic double helix model that illustrates the nature of DNA—the "precarious assemblage of molecules" which holds the secret of life.

Need to Know: Watson and Crick's discovery of the structure of DNA in 1953 posed a new set of questions about how to manipulate that chemical structure to the benefit of humanity.

"That Damned, Elusive Pimpernel"

Knowing the structure of DNA, scientists sought to find out how the molecular model actually transmitted characteristics. George Beadle and Edward Tatum proved that the action of a gene is to encode information to construct a protein, which in turn enables the specific form or function—the trait—within an organism.

Later, Watson and Crick discovered their "central

SUMMARY AND ANALYSIS

dogma," that RNA—ribonucleic acid, DNA's molecular cousin—is used as a "messenger molecule" to transmit copies of the DNA chain to build a new protein.

Need to Know: The "central dogma" of biological information is that DNA is transformed into RNA, a messenger molecule that constructs a new protein.

Regulation, Replication, Recombination

Experiments by the French biologist Jacques Monod proved that genes were not just passive carriers of traits, but were capable of changing the information they held to respond to changes in the environment, He did so by exposing the bacteria E. coli (Escherichia coli) to two sugars—glucose and lactose—and noting the genes' different reactions to each one. Monod showed that genes are able to regulate themselves, to replicate themselves, and also to recombine, creating new gene forms.

Need to Know: The "three R's" of gene physiology: genes can regulate themselves, replicate, and recombine to create new gene forms.

From Genes to Genesis

If proving the existence of the gene solved the problem of heredity, scientists still sought to learn how an entire organism grows out of a single cell. Embryologist Ed Lewis of Caltech concluded that certain genes (effectors) start this process by turning other genes on or off. By experimenting with inchworms, which have a finite number of genes, Lewis discovered that each gene has a specific function to perform within the organism, with some even "shutting off" to cause its death.

Genes, apart from being programmed with certain traits, were also affected by their proximity to, and interaction with, other genes. By way of analogy, the wood used to construct a ship contains certain intrinsic qualities; but it is the way the different sections are combined with the others that creates a floating vessel.

Need to Know: The way genes create a complex organism is determined not only by the traits they carry, but by their proximity to, and interaction with, other genes.

SUMMARY AND ANALYSIS

Part Three: "The Dreams of Geneticists"

Sequencing and Cloning of Genes (1970-2001)

"Crossing Over"

In 1973, biochemist Paul Berg created "recombinant DNA" for the first time by joining genetic material from different sources and using virus-bacteria hybrids as carriers. Shortly after, however, researchers Herb Boyer, Stanley Cohen, and Stan Falkow discovered a way to create genetic hybrids with bacterial genes—which were far less hazardous than the virus-bacteria hybrids—and reproduce them in large quantities in an incubator. This effectively ushered in "the birth of a new world," in which genes could be manipulated for therapeutic ends.

Need to Know: The creation of recombinant DNA and hybrids opened a world of possibilities for genetic manipulation.

The New Music

British biochemist Frederick Sanger learned how to read genes by "copying" a DNA sequence; rather than break down the DNA into its component parts, he slowed its reproductive process so he could observe the order in which the parts appeared. This brought science another step closer to altering genes. For decades, scientists had managed to do so only by bombarding genes with X-rays and producing mutants; by the late 1970s, however, gene cloning and gene sequencing (determining the order of DNA nucleotides in a gene) were a reality.

Need to Know: The capacity to alter genes created a divide in the science of biology, between the old guard who merely sought to describe and classify organisms, and the new biologists who studied ways to change them.

Einsteins on the Beach

Scientists who gathered at Asilomar, California, in 1975 drew up a scheme to rank the risks of biohazards of experiments with recombinant DNA and restrict research accordingly. The conference espoused self-regulation by scientists in order to avoid having restrictions imposed later by governments.

Need to Know: The Asilomar conference recommended self-governance by scientists in gene experiments, but did not address the ethical dilemmas raised by the nascent science of genetic engineering.

SUMMARY AND ANALYSIS

"Clone or Die"

During the 1970s, genetic science began to move inexorably into the realm of technology and, by extension, into business. Researcher Herb Boyer and venture capitalist Robert Swanson formed the genetic engineering company, Genentech, which in 1978 manufactured insulin from recombinant DNA.

Need to Know: The founding of Genentech and its success in creating recombinant insulin marked the start of the biotechnology industry.

Up until then, insulin was still being obtained by a crude and inefficient method which consisted of crushing pig and cow pancreases. Genentech patented the recombinant insulin and went on to invent many other genetic-based medicines.

Part Four: "The Proper Study of Mankind Is Man"
Human Genetics (1970-2005)

The Miseries of My Father

Late in life, Mukherjee's father fell from a rocking chair; this accident caused him to suffer from confusion, urinary incontinence, and difficulty in walking. He was diagnosed with Normal Pressure Hydrocephalus (NPH), a condition thought to be genetic, although it requires external circumstances to trigger the symptoms.

Need to Know: The experience of Mukherjee's father illustrates how genetic science must address human illness: by considering genetic structure, its variants, the environment, and chance, all of which influence human pathology.

The Birth of a Clinic

As genetic research in laboratories advanced, clinical studies were establishing the links between genetics and human illness. Victor McKusick founded the Moore Clinic in 1957 to study hereditary disorders; by 1998, he had discovered 12,000 gene variants linked to different medical disorders and traits.

Need to Know: The ability to detect genetic illnesses such as Down syndrome in fetuses created a "medical industry" in the 1970s, with the abortion of affected fetuses as one of its activities.

SUMMARY AND ANALYSIS

"Interfere, Interfere, Interfere"

Genetic testing on fetuses in the 1970s and the legalization of abortion was, in effect, a return to eugenics, which was now called neogenics, or newgenics. The difference was that in newgenics, genes were used as the basis for selection by individual choice.

Need to Know: In the "newgenics" that arose in the 1970s, the genetic testing and medical procedures were carried out by individual choice, not mandated by governments.

A Village of Dancers, an Atlas of Moles

By the 1970s, many diseases had been classified as genetic, but scientists were still unable to identify the specific genes that caused them. By compiling data on the affected families and using new mapping techniques—which pinpointed a gene's location through its linkage with other genes—researchers were able to identify the single gene that causes Huntington's disease. The same techniques were later used on cystic fibrosis. By the late 1990s, many genetic diseases could be detected through prenatal screening.

Need to Know: Scientists' ability to map genes

allowed them to detect genetic illnesses, opening a new horizon in man's ability to control human nature.

"To Get the Genome"

The successful mapping of disease-provoking genes gave impetus to a much larger project: mapping the entire human genome, with its more than 3 billion base pairs of DNA. Estimates of the cost and the time it would take were huge, but so were the potential rewards in diagnosing diseases, particularly cancer. The Human Genome Project got underway in 1989.

Need to Know: Starting in 1989, the Human Genome Project began to map the human genome, the entire "encyclopedia" of genetic information in the body.

The Geographers

Microbiologist Craig Venter decided to shortcut the process of mapping the human genome by examining only fragments of each gene. His company, Celera, entered into a race with the National Institutes of Health's Human Genome Project. In June 2000, the leaders of the two efforts announced jointly that they had made a "first survey" of the entire human genome.

Need to Know: In 2000, the Human Genome Project

SUMMARY AND ANALYSIS

and Celera announced they had made a "first survey" of the more than 3 billion base pairs of DNA in the human genome.

The Book of Man (in Twenty-Three Volumes)

This brief chapter summarizes the characteristics of the human genome in 23 bullet points and quotes, corresponding to the 23 pairs of chromosomes the genome contains. Perhaps the most singular characteristic of the human genome, compared to its animal or plant counterparts, is not the number of genes it contains, which is small compared to other organisms, but their complexity and diversity. The human genome has 20,687 genes; there are 32,000 in corn and more than 45,000 in wheat. However, the human gene network is marked by variety and by the sophisticated manner in which it operates.

Need to Know: The human genome, which is comprised of 23 pairs of chromosomes, is striking, not for the great number of genes it contains (wheat, for instance, contains more), but rather for its complex and diverse nature.

THE GENE

Part Five: Through the Looking Glass
The Genetics of Identity and Normalcy (2001-2005)

"So, We's the Same"

The complete map of the human genome has revealed the similarities, rather than the differences, between races. The genes that determine racial differences comprise only a small part of our genetic makeup, about 7%. The majority of genetic differences between people—in characteristics such as intelligence, personality, and temperament—occur within races, not between one race and another.

Genes themselves do not explain human diversity; that can only be understood by studying environment, culture, geography, and history.

Need to Know: The map of the human genome has shown that the majority of genetic differences exist between persons of the same race, not between one race and another.

The First Derivative of Identity

Genetics can determine some aspects of identity, but others depend on external factors. However, sex, defined as maleness or femaleness, is completely

SUMMARY AND ANALYSIS

genetic in origin. The presence of the Y chromosome determines maleness, but it must be activated by the SRY (sex-determining region Y) gene. When this doesn't occur, confusion about sexual identity results. During the 1970s, the idea that gender could be imposed externally ("nurture can overcome nature") gave rise to experiments in sexual reassignment that completely ignored genetic factors, to tragic results.

Need to Know: Genetics determine a person's sex as male or female, but gender identity is decided by a series of genetic and environmental factors.

The Last Mile

For decades, science had taken the view that personality traits, temperament, and sexual preference were all the results of environment. However, studies with identical twins who were separated at birth and raised in totally different environments showed striking similarities in all these areas and more. For example, a pair of identical twin brothers raised apart in the Midwest both grew up to be chain smokers, stock-car racing enthusiasts, and named their dogs Toy. The evidence, anecdotal and scientific, pointed to a genetic component of personality and gender identity.

Need to Know: Genes can point to propensities in

human behavior and personality, but they cannot determine outcomes—they cannot travel "the last mile."

The Hunger Winter

Survivors of the "Hunger Winter" of 1944 in Nazi-occupied Holland passed on the effects of that famine—heart disease, obesity, depression, anxiety—to their children and grandchildren. Research has shown that external events—whether traumatic or highly pleasurable—are recorded in cells by chemical markers that respond to various stimuli. The field of "epigenetics" studies changes in genes, which are inherited, but not caused, by changes in the DNA code.

Need to Know: Epigenetics analyzes the permanent effects that external events, unrelated to changes in DNA, can have on genes.

SUMMARY AND ANALYSIS

Part Six: Post-Genome
The Genetics of Fate and Future (2015-...)

The Future of the Future

Embryonic stem cells (ES) are contained in an embryo and can reproduce every cell in a living organism. They can also be extracted, genetically modified, and then reinserted into the organism. This knowledge gave rise to the first attempts at gene therapy during the 1990s. One of the first patients to undergo such a procedure, 18-year-old Jesse Elsinger, died as a result. Gene therapy was put on hold for nearly a decade.

Need to Know: The early experiments in gene therapy during the 1990s were unsuccessful and in some cases had tragic results, putting a stop to such procedures for years.

Genetic Diagnosis: "Previvors"

While gene therapy languished, progress continued in diagnosing genetic illnesses. Genetic diagnosis of ailments such as breast cancer or schizophrenia has created a new class of patients, called "previvors," who know that they can be hit with a life-threatening disease and must decide how to use that knowledge.

Need to Know: The ability to detect the presence of genetic illnesses has meant that some patients now must live with the knowledge that they may be struck by a debilitating or life-threatening disease.

Genetic Therapies: Post-human

Gene therapy comes in two distinct forms: alteration of a single gene to correct a deficiency, a change that will last only one generation; or permanent alteration of reproductive genes, called "germ-line editing." A group of scientists in 2015 called for a moratorium on the use of gene-alteration and gene editing technologies "in a clinical setting," particularly in human embryo stem cells. In a statement, they expressed concern over the possibility "of initiating a 'slippery slope' from disease-curing applications towards uses with less compelling or even troubling implications."

Meanwhile, Chinese scientists attempted unsuccessfully to genetically modify a human embryo. Chinese experimentation will continue, however, and the creation of the first human from a genetically altered embryo is not far off.

Need to Know: Gene therapy can alter a single gene to correct a deficiency, or alter the reproductive genes, effecting a permanent change in subsequent generations of an organism.

Epilogue

Our capacity to identify genetic illnesses and cure them has marked a huge advance in science, while at the same time posing a series of moral and ethical questions to the whole of society. According to the author, the influence of genes on humanity is "richer, deeper, and more unnerving than we had imagined"; managing them, he says, "will be the ultimate test of discernment for our species."

Need to Know: Despite all the advances in human biology, including the mapping of the human genome, our knowledge of human genes remains very limited.

Timeline

350 BC: Aristotle theorizes that information about hereditary traits in humans is passed on from generation to generation in the form of messages.

1859: Publication of Charles Darwin's groundbreaking work, *On the Origin of Species.*

1865: Augustinian monk Gregor Mendel discovers that hereditary traits are contained in indivisible units. These would later be termed "genes."

1869: Francis Galton invents the term *eugenics* to describe the efforts to improve biological and psychological traits by controlling human breeding.

SUMMARY AND ANALYSIS

1900-1909: Mendel's work finally comes to the attention of scientists; the term *gene* is coined by botanist Wilhelm Johanssen in 1909 to describe the individual unit of heredity.

1933-1945: In Nazi Germany, the state imposes practices of "racial hygiene." Genetic experiments are carried out by Josef Mengele on prisoners in concentration camps.

1944: Oswald Avery discovers that DNA carries the genetic material of a cell.

1953: James Watson and Francis Crick, with the aid of Rosalind Franklin and Maurice Wilkins, devise their now-emblematic model of DNA.

1973: Paul Berg, Herb Boyer, and Stanley Cohen create the first molecule of recombinant DNA, which brings together material from multiple genetic sources.

1975: Scientists meeting at the Asilomar conference agree to place a moratorium on experiments with recombinant DNA, until all the potential dangers are assessed.

1976: Cancer is proven to be caused by multiple genetic mutations.

1998: Human embryonic stem cells are obtained from embryos.

1999: Jesse Gelsinger dies while undergoing an experimental attempt at gene therapy. His death effectively puts gene therapy on hold for more than a decade.

2000: The Human Genome Project publishes the draft sequence of the human genome.

2003: Sequencing of the 3 billion base pairs that constitute the human genome is successfully completed by the International Human Genome Project.

2012: Jennifer Doudna and Emmanuelle Charpentier discover a method for reliable, efficient alteration of human genes. Genome editing, or genomic surgery, becomes possible.

2015: Doudna joins other scientists in calling for a moratorium on gene editing and gene altering in a clinical environment. Scientists in China attempt to modify the genome of a human embryo, without success, but announce their intention to continue working toward this goal.

Cast of Characters

Aristotle: Ancient Greek philosopher and scientist who theorized that human traits were passed on from generation to generation as messages.

Paul Berg: Biochemist at Stanford University who, in collaboration with Herb Boyer and Stanley Cohen, in 1975 discovered recombinant DNA, which brings together material from multiple genetic sources.

Herb Boyer: Scientist who collaborated with Paul Berg and went on to found Genentech, the biotechnology company that used recombinant DNA to produce synthetic insulin and other medicines.

SUMMARY AND ANALYSIS

Carrie Buck: West Virginia woman who was legally sterilized after being diagnosed as an "imbecile" during the rise of the eugenics movement in the 1920s.

Charles Darwin: British naturalist whose seminal work *On the Origin of Species* proposed a theory of evolution by "natural selection." However, Darwin recognized that his work did not provide an adequate explanation for heredity.

Hugo de Vries: Dutch botanist and geneticist who confirmed Mendel's theories and discovered the process of mutation in plant and animal life.

Francis Galton: Charles Darwin's cousin, proponent of the practice of eugenics, and author of the phrase "nature versus nurture."

Frederick Griffith: English bacteriologist who made the first successful gene transformation.

Gregor Johann Mendel: Eighteenth-century Augustinian monk whose work breeding hybrid strains of pea plants confirmed the existence of genes as indivisible units of heredity.

THE GENE

Wilhelm Johannsen: Danish botanist who coined the term "gene" to describe the individual unit of heredity.

Josef Mengele: The "Angel of Death" doctor who performed genetic experiments on twins and on deformed individuals held in concentration camps in Nazi Germany.

Frederick Sanger: British biochemist and two-time Nobel laureate whose work on how to read the chemical composition of the genes contained in DNA revolutionized the science of genetics.

Nettie Stevens: Biologist at Columbia University who discovered that the male sex was determined by the presence of the Y chromosome in embryos.

James Watson and Francis Crick: Researchers at King's College, London who, along with Maurice Wilkins and Rosalind Franklin, developed the famous double-helix model of DNA.

Nancy Wexler: Psychologist whose research, using modern gene-mapping techniques, led to the discovery of the gene that causes Huntington's disease.

Direct Quotes and Analysis

"Memories sharpen the past; it is reality that decays."

The author makes this comment during a visit to his father's boyhood home, which now appears smaller and shabbier than it did years before. He is referring to the phenomenon of remembering details—in this case, details mostly borrowed from his father's stories—in an outsized way. Having viewed this home through the eyes of a child, his father had described the place as being larger both in reality and in meaning than it appears to them when they visit as adults.

SUMMARY AND ANALYSIS

"There is no such thing as perfection, only the relentless, thirsty matching of an organism to its environment. That is the engine that drives evolution."

The debate over "nature versus nurture"—the influence of genes, as opposed to environment—has been central to the history of genetics. The reality is that nature and nurture work in concert to create the entirety of evolutionary history. Individuals are impacted not only by their genetic makeup, but also by the effects of their surroundings.

"Like musicians, like mathematicians—like elite athletes—scientists peak early and dwindle fast. It isn't creativity that fades, but stamina: science is an endurance sport."

While the author's assertion may be true in general, there are, of course, exceptions. But these late-in-life discoveries are rare, as thousands of failed experiments must be put in the past before a great breakthrough; it takes a certain tenacity to develop transformative research. Mukherjee observes that university professor Oswald Avery was 66 years old in 1944, when he proved that DNA was the carrier of genetic information.

THE GENE

"It is the impulse of science to try to understand nature, and the impulse of technology to try to manipulate it."

The Asilomar Conference in California marked a sea change in the way scientists thought about genes and the way their research could be used going forward. The invention of recombinant DNA in 1975 had given them the ability to manipulate genes—and therefore to take that knowledge into the realm of technology.

"The problem of racial discrimination . . . is not the inference of a person's race from their genetic characteristics. It is quite the opposite: it is the inference of a person's characteristics from their race."

While experiments in eugenics were largely based on the idea that the characteristics of "inferior" racial groups could be erased on a genetic level, recent revelations in the field put to rest the idea of there being significant genetic differences between races. Recent studies show that the vast proportion of genetic diversity (85–90%) occurs within races, and only 7% between racial groups.

Trivia

1. The phrase "survival of the fittest" was not coined by Charles Darwin, the man most associated with it, but by the Malthusian economist Herbert Spencer.

2. Gregor Mendel, who correctly identified the gene as the basic unit of heredity, sent his work to Charles Darwin, but there is no evidence that Darwin ever read it. Mendel's research would have provided the "missing link" in Darwin's theory of evolution.

3. The first survey of the human genome was hailed as a cooperative effort between a private company,

SUMMARY AND ANALYSIS

Celera, and the National Institutes for Health. But really, the two were archrivals in the race to map the human genome, and were persuaded to present their findings jointly after some prodding by President Bill Clinton.

4. One of the stars of genetic science in the early 1930s was Hermann Muller, who proved that the genes of fruit flies could be altered by exposing them to low doses of radiation. Muller, who later won the Nobel Prize for his work, was a committed socialist who was investigated by the FBI for his political activities. Longing to live in a society where socialism and science could work hand-in-hand, he left the United States for Germany in 1932—ironically, on the eve of the Nazis' rise to power.

5. Although the eugenics movement was widely discredited by the horrors of Nazi Germany, it did not disappear altogether. In the 1990s, under President Alberto Fujimori, the Peruvian government sterilized thousands of mostly poor, indigenous men and women. The rationale was that poverty could be reduced by lowering the birth rate. Fujimori, now in jail for corruption and human rights abuses, has denied that anyone was forcibly sterilized during his administration.

6. Early experiments with embryonic stem cells in the 1990s led to the first "transgenic" animals, whose genes had been permanently altered, with some curious results. Mice injected with growth hormones grew to twice their normal size, while a mouse that was given genes from a jellyfish glowed in the dark when exposed to a blue light. However, it was soon discovered that human genes were not so easily manipulated.

7. The publication of *The Gene* in May 2016 was preceded by an article by Mukherjee in the *New Yorker* magazine regarding epigenetics, the field that studies how environmental factors can affect genes separately from the DNA sequence. The article, based on the book, raised a wave of criticism from experts who said the science in it was inaccurate and misleading. Specifically, they said Mukherjee had minimized the importance of transcription factors—proteins that turn genes on and off—in order to concentrate on less important aspects of the process. Mukherjee issued a rebuttal to his critics, defending both the article and the book. But in July 2016, Scribner's announced that changes and clarifications had been made in the fifth printing of *The Gene,* in response to the issues raised by the controversy.

What's That Word?

Chromosome: A threadlike structure within a cell, comprised of proteins and DNA, that holds genetic information in the form of genes.

DNA: Deoxyribonucleic acid, the main constituent of chromosomes and the carrier of genetic information.

Enzyme: A substance produced by a living organism that acts as a catalyst to provoke a specific biochemical reaction.

Epigenetics: The study of heritable changes in gene function that do not involve changes in the DNA sequence.

SUMMARY AND ANALYSIS

Eugenics: A practice that seeks to improve the hereditary qualities of a race or breed, as by controlling human mating.

Gene: The basic unit of heredity in biology, which is transferred from parent to offspring and determines some trait of the offspring. A gene is a specific set of nucleotides in DNA or RNA that is usually located on a chromosome.

Genome: The complete set of all genetic information present in a cell or organism; the full DNA sequence of an organism.

Mutation: A change in the chemical structure of the DNA of an organism, which can affect its structure and functioning. The resulting variant forms may be transmitted to subsequent generations.

Protein: A chemical comprised of a long chain of amino acid, which is a fundamental part of any living organism.

RNA: Ribonucleic acid, a chemical that serves as a "messenger" for a gene to be translated into a protein.

Stem cell: A cell that can reproduce itself and also give rise to other cells, through differentiation.

Transgenic: An organism whose genes have been modified, e.g., transgenic animals.

Transformation: The transfer of genetic material from one organism to another, without reproduction.

Critical Response

"[*The Gene*] is destined to soar into the firmament of the year's must-reads, to win accolades and well-deserved prizes, and to set a new standard for lyrical science writing." —Abigail Zuger, MD,
The New York Times

"*The Gene* is a frank celebration of progress—the immense and extraordinarily rapid increase of our knowledge of what genes are and how they work—but Mukherjee is concerned about what that knowledge is doing." —Steven Shapin,
The Guardian

SUMMARY AND ANALYSIS

"Mukherjee leaves you feeling as though you've just aced a college course for which you'd been afraid to register—and enjoyed every minute of it."

—Andrew Solomon,
The Washington Post

"*The Gene*'s dominant traits are historical breadth, clinical compassion, and Mukherjee's characteristic graceful style." —Nathaniel Comfort,
The Atlantic Monthly

"Sobering, humbling, and extraordinarily rich reading from a wise and gifted writer who sees how far we have come—but how much farther we have to go to understand our human nature and destiny."

—*Kirkus Reviews*

About Siddhartha Mukherjee

Siddhartha Mukherjee is best known as the author of *The Emperor of All Maladies: A Biography of Cancer*, which won the 2010 Pulitzer Prize for general nonfiction and was made into a three-part documentary film by director Ken Burns. Mukherjee is currently an assistant professor of medicine at Columbia University in New York, where he studies blood cancers, and a staff physician at New York-Presbyterian Hospital.

Born in Calcutta, Mukherjee is a Rhodes scholar who graduated from Stanford University, Oxford

SUMMARY AND ANALYSIS

University, and the Harvard Medical School. His work has been published in the *New York Times*, the *New Yorker*, the *New Republic*, *Nature*, *Cell*, and the *New England Journal of Medicine*.

In May 2016, Mukherjee announced that he has cofounded a new biotech company, Vor BioPharma, which will concentrate on using CAR-T cells (genetically modified white blood cells) to recognize and kill cancer cells in leukemia patients.

For Your Information

Online

"'The Gene: An Intimate History' by Siddhartha Mukherjee Decodes DNA, Genetics." ChicagoTribune.com

"Genes Are Overrated." TheAtlantic.com

"Gene Tests Identify Breast Cancer Patients Who Can Skip Chemotherapy, Study Says." NYTimes.com

"The Power of Genes, And The Line Between Biology And Destiny." NPR.com

"Pulitzer Prize–Winning Writer and Physician Introduces His New Book *The Gene: An Intimate History*." CharlieRose.com

"Siddhartha Mukherjee Discusses *The Gene*." C-SPAN.com

SUMMARY AND ANALYSIS

Books

The Age of Genomes: Tales From the Front Lines of Genetic Medicine by Steven Moore Lipkin, MD with Jon R. Luoma

Born That Way: Genes, Behavior, Personality by William Wright

The Double Helix: A Personal Account of the Discovery of DNA by James D. Watson

Epigenetics: The Ultimate Mystery of Inheritance by Richard C. Francis

Genome: The Story of the Most Astonishing Scientific Adventure of our Time—The Attempt to Map All the Genes in the Human Body by Jerry E. Bishop and Michael Waldholz

On the Origin of Species by Means of Natural Selection by Charles Darwin

The Selfish Gene by Richard Dawkins

A Troublesome Inheritance: Genes, Race, and Human History by Nicholas Wade

Other Books by Siddhartha Mukherjee

The Emperor of All Maladies: A Biography of Cancer

The Laws of Medicine: Field Notes from an Uncertain Science

Bibliography

Comfort, Nathaniel. "Genes Are Overrated," *Atlantic Monthly*, June 2016.

Herper, Matthew. "Siddhartha Mukherjee, Author of Bestselling Cancer Book, Starts Biotech Company and Answers Criticism," *Forbes*, May 9, 2016.

Kyriacou, Charalambos P. "'The Gene: An Intimate History,' by Siddhartha Mukherjee," *Times Higher Education*, July 7, 2016.

Maloney, Jennifer. "Publisher Tweaks 'Gene' Book After New Yorker Article Uproar," *Wall Street Journal*, July 28, 2016.

Ritchie, Stuart. "How Siddhartha Mukherjee gets it wrong on IQ, Sexuality, and Epigenetics." *The Spectator*, June 28, 2016.

SUMMARY AND ANALYSIS

Shapin, Steven. "'The Gene' by Siddhartha Mukherjee (review)," *Guardian*, May 25, 2016.

Ward, Andrew. "The Gene: An Intimate History" (review), *Financial Times*, June 10, 2016.

Woolston, Chris. "Researcher Under Fire for *New Yorker* Epigenetics Article," *Nature*, May 19, 2016.

WORTH BOOKS
SMART SUMMARIES

So much to read, so little time?

Explore summaries of bestselling fiction and essential nonfiction books on a variety of subjects, including business, history, science, lifestyle, and much more.

Visit the store at
www.ebookstore.worthbooks.com

MORE SMART SUMMARIES
FROM WORTH BOOKS

POPULAR SCIENCE

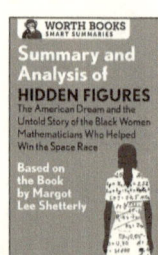

Find a full list of our authors and
titles at www.openroadmedia.com

FOLLOW US
@OpenRoadMedia

www.ingramcontent.com/pod-product-compliance
Lightning Source LLC
Chambersburg PA
CBHW060342080526
44584CB00013B/886